THIS IS NUMBER
ONE HUNDRED AND FORTY ONE
IN THE SECOND NUMBERED SERIES
OF THE MIEGUNYAH VOLUMES
MADE POSSIBLE BY THE
MIEGUNYAH FUND
ESTABLISHED BY BEQUESTS
UNDER THE WILLS OF
SIR RUSSELL AND LADY GRIMWADE.

'MIEGUNYAH' WAS THE HOME OF
MAB AND RUSSELL GRIMWADE
FROM 1911 TO 1955.

A Hostile Beauty

A Hostile Beauty

LIFE ON MACQUARIE ISLAND

Alistair Dermer

and Danielle Wood

THE
MIEGUNYAH
PRESS

To my children Asha, Clay and Zavier, may this book inspire you to follow your dreams. To seek out all the unique, wild, degraded, raw and peaceful parts of our planet. In your own way I know you will continue to have a positive influence on humanity and our globe.

THE MIEGUNYAH PRESS
An imprint of Melbourne University Publishing Limited
187 Grattan Street, Carlton, Victoria 3053, Australia
mup-info@unimelb.edu.au
www.mup.com.au

First published 2011
Text © Danielle Wood, 2011
Photography © Alistair Dermer, 2011
Design and typography © Melbourne University Publishing
Limited, 2011

Designed by Pfisterer+Freeman
Printed in Singapore by Imago

National Library of Australia Cataloguing-in-Publication entry:

Dermer, Alistair.

A hostile beauty: life on Macquarie Island / Alistair Dermer and Danielle Wood.

9780522855043 (hbk)

Macquarie Island (Tas.)—Description and travel.
Macquarie Island (Tas.)—Pictorial works.

Wood, Danielle.

919.48

CONTENTS

FOREWORD

Just as our Earth precipitated out of galactic gases billions of years ago and uniquely gave rise to life as we know it, so Macquarie Island uniquely arises from an ongoing collision of oceanic plates that began less than a million years ago. It has become a remarkable wildlife platform in a vast expanse of ocean. In 2001, young Australian photographer and environmentalist Alistair Dermer fulfilled a dream by spending his summer in a hut on a beach on the island. Out of that dream and summer comes this splendid book.

Alistair's photographs are a song to our kinship with nature. The delight with his sojourn in one of the world's most remote but lively places shows in every picture. That delight is there in the caressing lambs-ear foliage of Macquarie Island, the imperious squadrons of orange-collared penguins, the fat seal pups with their endearing eyelashes, the droplets of drizzle on the infinity of fine filaments of the edible native cabbage and the lush, spongy landscapes giving way to the kelp-strewn, rocky shores.

It wasn't always so. Macquarie Island's first century of human history is ugly: shipwrecks, nail-studded clubs, seal slaughters, iron penguin-digester vats and introduced cats, rats and rabbits ravaging the ecosystem. It was followed by a more hopeful century of visitors who came home from the island entranced by its beauty rather than its vulnerability to exploitation, and determined to work for its preservation. Today, the island Alistair pursued and caught with a camera rather than a club is one of the world's great wildlife refuges. Its wildness is rarer than almost anywhere on the planet and its herds of wildlife numerically rival those of Africa and Alaska.

The background miracle of this book is that such a stunning natural domain has revived through human choice. Alistair's vivacious photographs of Macquarie Island also illuminate the flame of hope that the eco-tragedy we live in beyond Macquarie Island can be reversed and life on the rest of the Earth, in all its splendour, can equally be reassured in the years ahead.

Bob Brown

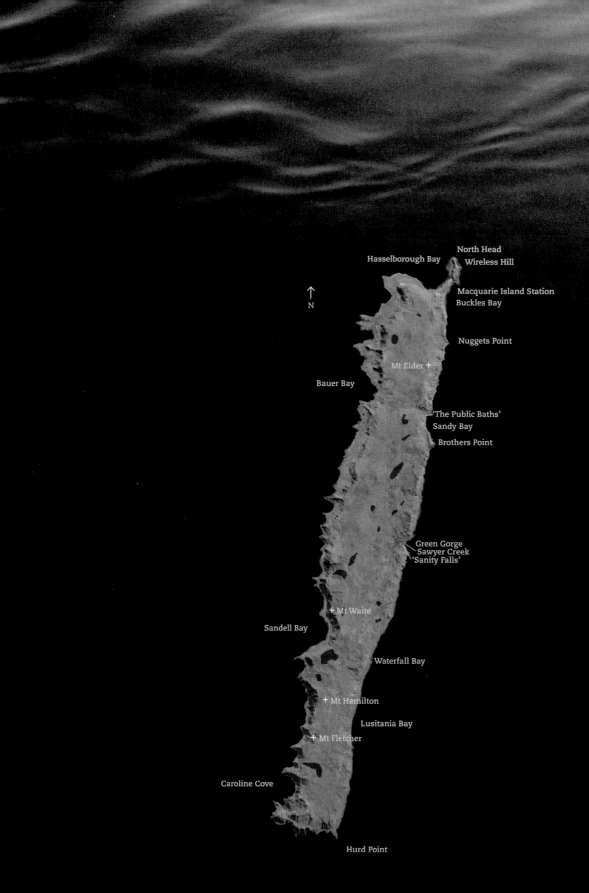

North Head
Wireless Hill
Hasselborough Bay

Macquarie Island Station
Buckles Bay

Nuggets Point

Mt Elder +

Bauer Bay

'The Public Baths'
Sandy Bay

Brothers Point

Green Gorge
Sawyer Creek
'Sanity Falls'

+ Mt Waite

Sandell Bay

Waterfall Bay

+ Mt Hamilton

Lusitania Bay

+ Mt Fletcher

Caroline Cove

Hurd Point

Macquarie Island

0 ————————— 5km

PREFACE

A gentle Southern Ocean, drizzle and a strong sea mist were the conditions as I peered out of my porthole on 1 October 2001. What I glimpsed was a jagged bit of land coming straight out of the ocean and eerie sea fog. Scattered around were rocks that looked like they had claimed many ships. After almost three days at sea, the *Aurora Australis* had delivered me to what was to be my subantarctic home for the next six months.

This was the beginning of what was to be a unique time in my life, an opportunity to maintain a healthy level of sustained reflection. It was a time that forced me in many ways to release the trivial expectations and perceptions of the civilised world and see nature at its best or, at least what cannot be argued, its most raw. Newborn chicks are stolen, lost, squashed, frozen. Elephant seal bulls, their immense bulk pumped up with testosterone, squash everything in their path when they charge towards an intruder threatening their dominance over the harem. I also witnessed the beauty of nature in a way I had not before. One of my most treasured memories is of a wandering albatross landing near me, and then moving closer until it was sitting against me. It was not only a unique and privileged wildlife

Opposite: Humans are not permitted near the wandering albatross, but on this day, this bird came to me. It was an amazing experience but also sad because in all likelihood the bird was lonely, its partner having never returned.

experience but also a reminder that these birds wait each year for their life partner to return, often in vain. They have met their death more often than not from impacts associated with human activities.

My trip to Macquarie Island was one that manifested from a childhood dream. I remember repeatedly looking at a book that featured Macquarie Island. Other wonderful places like the Galapagos Islands and Antarctica dominated the pages, but it was Macquarie Island that inspired me. I would turn the pages and lose myself in the images, and feel somehow stirred, trying to imagine if such a place really existed and if people could ever visit. As my life experiences mounted up and my sensitivity to the natural world deepened, this sense of a calling and awareness was only to increase. It was only a matter of time before I would get to the mysterious island.

I made several attempts to get to the island and eventually made it there through a friend from university who was studying the effects of human impact on sea birds, mostly royal penguins. Luckily for me, I had the skills for the job of field and technical assistant. After receiving the phone call informing me that I had been chosen for a summer job on Macquarie Island, I was ecstatic. I borrowed money and sold whatever I could to enable me to buy a few extra lenses for my old but trusty Nikkormat, and a second manual Nikon (FE model), a case to keep them dry when boating, fifty rolls of film and a sturdy tripod. My intention was to return with fifty good photos, from which I would select the better ones and have a small exhibition. I hoped this might inspire someone else to follow their dreams. I often reflect and feel that my photographs are, in themselves, sustained reflections.

Soon after my return from Macquarie Island, I showed a selection of photos to a friend of a friend, who suggested I make a book. The pattern of events that followed took the usual shape we experience in life—a few ups and downs—but, to my amazement and complete satisfaction, this book is in front of you.

I hope you find wonder in these images.
Alistair Dermer

'like a
signpost
pointing
the way to
the frozen
lands…'

'DEFINED BY ITS PROFOUND AND POWERFUL REMOTENESS'

Macquarie Island is a sliver of unexpected rock in the vast Southern Ocean. It is wild, beautiful, wind chilled and wave bitten—defined by its profound and powerful remoteness. An island born unto itself alone, Macquarie Island has never, in its geologically short history, touched another landmass.

At 54° S, Macquarie Island is almost a halfway point between Tasmania's southern coast and the Antarctic ice; its nearest neighbours are the Auckland Islands, more than 600 kilometres to the east. Thirty-four kilometres long and up to five kilometres wide, the steep and slender island broke the ocean's surface sea some 600 000 or more years ago. The tiny tip of a submerged section of the Macquarie Ridge, the island was—and still is—pushed upwards by the collision of the Pacific and Australian tectonic plates.

With its World Heritage listing, its status as a UNESCO Biosphere Reserve, and with its immediate waters forming a large Marine Protected Area, Macquarie Island has been acknowledged as a place of global significance. The World Heritage listing rests, in part, on the island's unique geology. Here, as nowhere else on Earth, perfectly preserved rocks from up to 6 kilometres below the ocean floor are exposed above sea level, enabling geologists to study oceanic crust

Opposite: King penguins assess the thundering easterly swells. Eventually they accept the only way to get to the feeding ground is via a battering.

Mount Martin: One of the
many plateau lakes after
a night of snow.

'DENSE CRECHES OF ELEPHANT SEAL PUPS THAT SEEM AS MUCH A PART OF THE LANDSCAPE AS THE ROCKS THAT SURROUND THEM'

formations that are usually well beyond the reach of drilling technology. The island is considered by the World Heritage Committee to be both 'an outstanding example representing major stages of the Earth's history' and a place of 'outstanding natural beauty and aesthetic importance'.

When Macquarie Island first emerged, it would have been nothing more than a minute rocky outcrop surrounded by millions of square kilometres of stormy ocean. Only thousands of years after the island's emergence would sea birds or seals be able to breed there, safe from the surging waves of the Southern Ocean. Wind and waves brought the ancestors of the island's indigenous flora and fauna to rest on its shores. But simply reaching the island gave no guarantees of establishment. Successful colonisation of this speck of land was both erratic and fortuitous.

Today, the island's 120 square kilometres of rock and vegetation give life to vast numbers of sea birds and seals. For the southern elephant seal, three species of fur seal, four species of penguin, four species of albatross and numerous other bird species, the island's shores are a place to moult and rest and breed. What the island lacks in species diversity it makes up for in sheer biomass. Colonies of penguins stretch the length of beaches and, in springtime, there appear dense

'A SINGLE DAY MAY CONTAIN FOG, DRIZZLE, HAIL, SNOW AND WINDOWS OF BRILLIANT SUNSHINE'

creches of elephant seal pups that seem as much a part of the landscape as the rocks that surround them.

This island is a place of textures, of gritty, black-sand beaches and leathery convolutions of olive-green kelp, of velvety sage-coloured herb fields and coastal terraces of quaking bog. On many of the island's peaks, mosses and lichens are ribbed with bare rock—an environment that botanists call 'feldmark'. Between the scalloped coves and sea-stacks of the island's wild west coast and its comparatively sheltered east coast are green-pelted slopes and a highland plateau dotted with wind-rippled tarns.

Lying in the path of the Furious Fifties, the island is characterised by oceanic weather that is at once energetically changeable and remarkably stable. Although a single day may contain fog, drizzle, hail, snow and windows of brilliant sun-shine, temperatures vary only a little during the day, and between the winter and summer months. Rain falls on most days of the year. When snow falls—and sometimes it falls as low as the island's beaches—it is quick to melt away even from the highest peaks. Wind is ubiquitous. To be on the island is to be buffeted and blustered by the prevailing westerlies.

Opposite: Waterfall Bay Hut. This hut provided a peaceful getaway for me only a couple of hours walk from Green Gorge. To awake to snow on the rocky beach was a special treat.

'an atmosphere oppressive'

Writing in the *Adelaide Mail* in 1947, journalist Osmar White imagined the qualities necessary for holding down the proposed position of Macquarie Island caretaker.

WANTED: Reliable man, trained in meteorology, young and in robust health, able to endure one of the world's most trying climates, must be fond of animals, and not afraid of his own company. The man—or men—will have to have temperamental as well as scientific qualifications. They will have to live in a land of almost perpetual fog and rain, where the mean annual temperature is just under 40[°F, 4° C]. They will have to endure the perpetual, monotonous thunder of mountainous surf on narrow shingle beaches beneath cliffs, the screaming of countless millions of sea birds. They will have to bear with equanimity the impertinence of sea lions who make a habit of visiting camps at night, the discomfort of having—literally—not a square yard of dry ground to walk on. And above all, they will have to contend with an atmosphere oppressive, with an isolation, a savagery, that no continental desert knows.

Green Gorge hut, where I lived for six months.

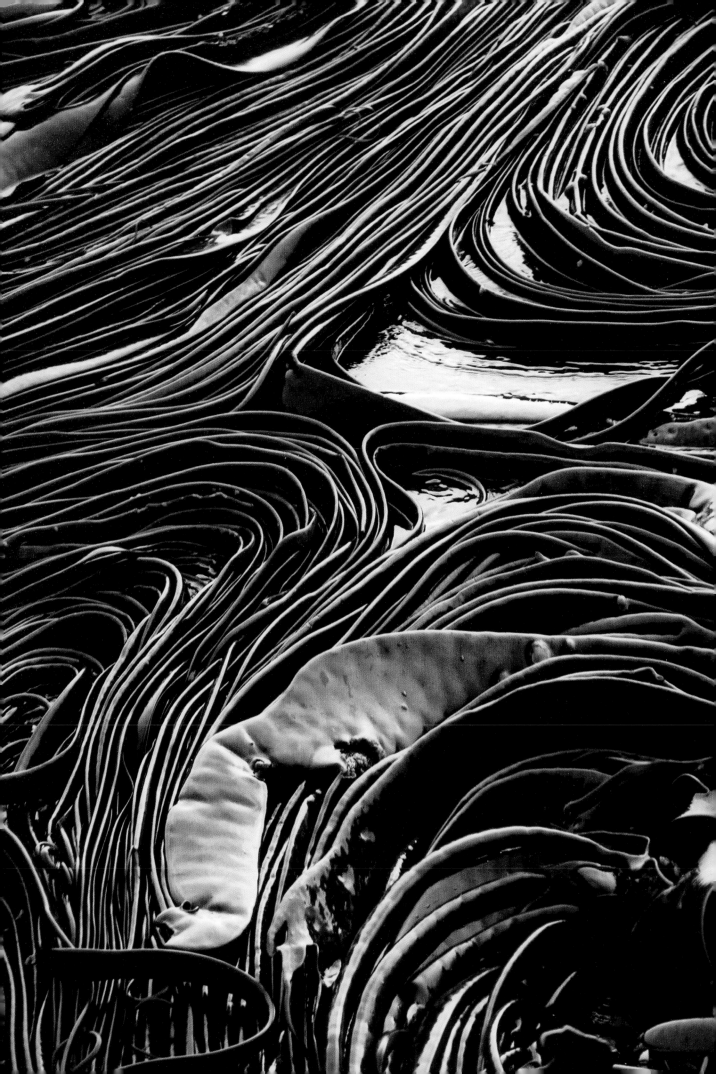

The island is not a human place, but human activity has left its inevitable footprint on the delicate subantarctic ecosystem. During the nineteenth century, people came to the island primarily to kill its wildlife. The skins of fur seals, the blubber of elephant seals and the oil from the fat, skin and bones of penguins were the raw products that enticed merchants to send their ships south. In the decades after the slaughter stopped, penguins and most of the seal species rebuilt their populations, but the devastating legacy of this first chapter of human visitation remained in the form of cats, rabbits, rats and mice.

Today, the island is visited not in the name of exploitation but in a spirit of awe. The few scientists, naturalists, expeditioners, artists and tourists who make the voyage south do so, in the most part, out of a desire to learn, to bear witness and to protect.

Opposite: Young elephant seals play fight in the shallow waters of Hasselborough Bay, only about 100 metres from the main station.

Following pages: Looking south from North Head over the station.

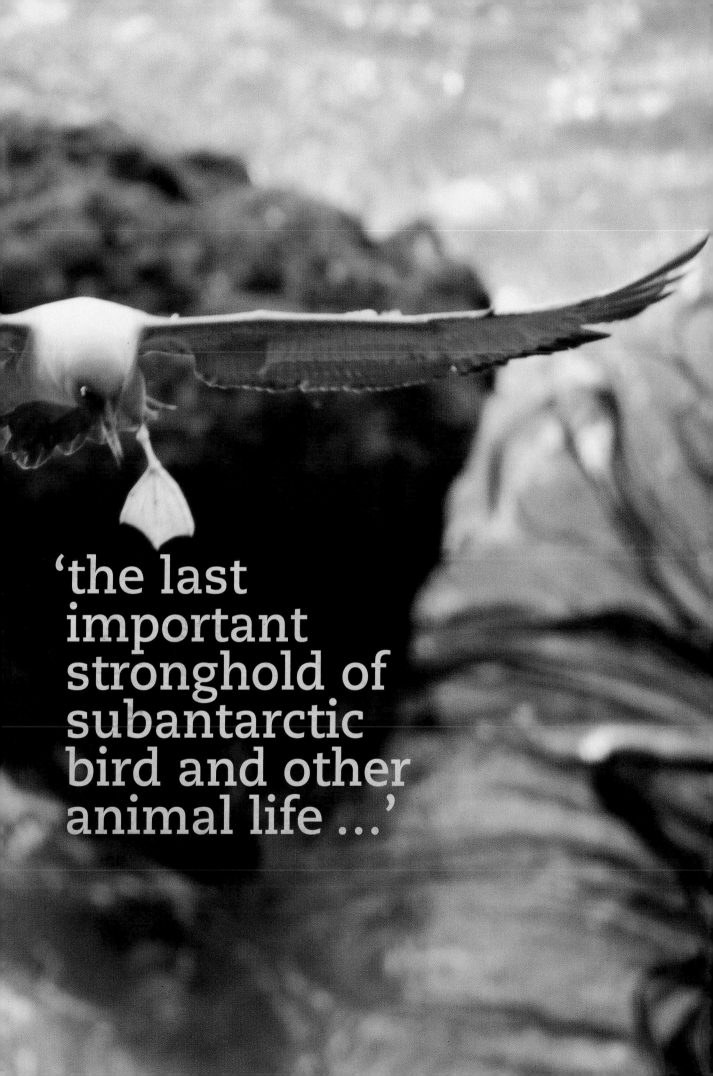

'the last
important
stronghold of
subantarctic
bird and other
animal life ...'

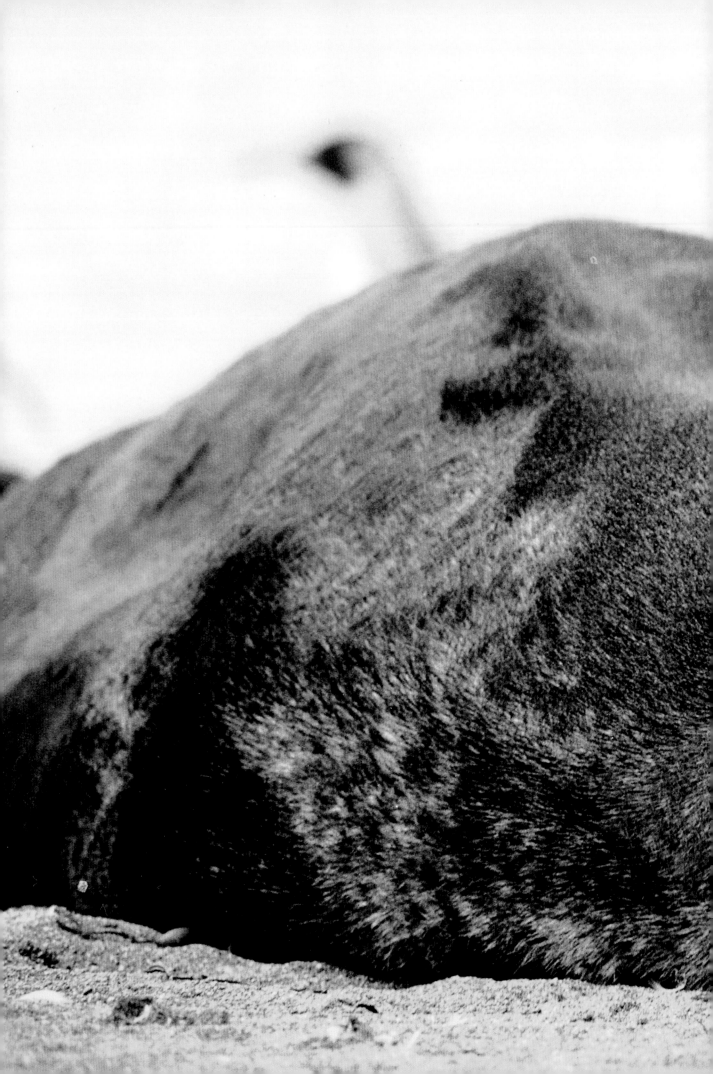

'THE TANG OF SALT AND KELP, THE FUG OF ANIMAL'

A Macquarie Island beach in early summer: the soundtrack is the crackle of waves withdrawing over grey shingle, the collective whirring of king penguins with their orange throats stretched skywards, and the overhead calling of ever-watchful skuas. The scent is the tang of salt and kelp, the fug of animal. Elephant-seal weaners play in the shallows, practising rearing up on their tails, mouths open wide in mock threat, then comically topple into the foam. A bull seal comes in from the sea, lumping his bulk up the beach, careless of all in his path. A gentle gentoo penguin—black head spattered with a constellation of white spots—watches from the tussocks.

For millions of individual marine birds and mammals that live and eat in the vast Southern Ocean, Macquarie Island is home base. Even if they might spend only short periods of their lives ashore, this sliver of rock is vital to their survival. It is a place for fur and elephant seals to moult, give birth and raise young, a place for penguins, albatrosses, petrels and other sea birds to nest and to fledge their chicks. An estimated one-seventh of the world's southern elephant seals call this island home. Approximately 850 000 breeding pairs of the endemic royal penguin mass here, forming one of the greatest concentrations of a species of sea bird in the world. King and

Previous pages: Flippers of male (left) and female (right) elephant seals.

Opposite: A female fur seal, bearing scars from the mating season, finds herself in the middle of a disapproving colony of royal penguins.

'ON LAND, THESE MASSIVE ANIMALS ARE NOISY, CHAOTIC AND CLUMSY'

royal penguin colonies blacken entire beaches with their sheer numbers.

The southern elephant seal, once known as the 'sea elephant', takes its name from the male seal's large proboscis, an appendage that becomes increasingly more elephantine when a bull is enraged. On land, these massive animals are noisy, chaotic and clumsy. Dominant bull seals, known as 'beach masters', police their harems of up to eighty cows with vigour. Indeed, they devote their existence to maintaining exclusive access to their females, and can be seen crushing their own offspring as they hump themselves over the beaches to see off competitors who hover around the edges of the harem, waiting for an opportunity to mate. At sea, where these seals spend most of their lives, they are quite different creatures. They become streamlined swimming machines—probably the best and most highly developed diving mammals in the world. Elephant seals are capable of diving to depths of 1000 metres, a feat requiring them to slow their heart rates to an extraordinary seven or eight beats per minute to shut off blood flow to every part of their body except their brains.

The elephant and fur seals that occur on Macquarie Island have developed a cunning biological plan to accomplish both birth and

Opposite: Fat and bored, this elephant seal bull will wait out the mating season on the sidelines while a bigger bull enjoys dominance over the harem.

mating activities in the short springtime period they spend ashore. Soon after the females give birth, the animals mate, but implantation of the fertilised egg is delayed for several months until it is time for the gestation of the next springtime pup to begin in earnest.

In contrast to the elephant seals, the fur seals are playful, mobile and almost freakishly dog-like. These agile creatures can be seen clustered, in springtime, on the island's northerly beaches, or porpoising in pods along the coastline. Three species occur here—subantarctic, Antarctic and New Zealand fur seals—and, although it was believed New Zealand fur seals only rarely bred on the island, recent research has shown a higher than expected rate of interbreeding between all species. Male subantarctic fur seals are striking with their blond chests, dark chocolate-coloured heads, flat-top hairdos and distinctive barking calls. Their Antarctic counterparts are bigger and shaggier, with long and droopy whiskers and a tendency to huff and puff when vocalising. In the fur-seal colonies, the breeding season is a time of frantic, and sometimes violent, activity. In fights over mating rights, competing males lash out at each other with their sharp teeth, trying to wound their opponent at the vulnerable junction between body and flipper.

Opposite: A female fur seal tagged for research. Researchers monitor the seals to understand population dynamics and measure any future declines against human activity.

'Although I had heard so often of the great quantity of birds on the uninhabited islands', wrote a sailor visiting Macquarie Island in 1840, 'I was not prepared to see them in such myriads as here. The whole sides of the rugged hills were literally covered with them'. Estimates of the total number of birds breeding on the island are close to four million, and the majority of these are penguins.

Four penguin species occur on the island, the most numerous being the curious and vivacious royals, with their pale pudgy feet and drooping yellow crests. Largest are the king penguins, which congregate in a vast rookery in Lusitania Bay in the island's south. These sleek birds with their immaculate black, white and orange livery are the world's second-largest penguins after the emperors of the Antarctic continent. The smallest of the Macquarie Island penguins are the aggressive rockhoppers, which make quite an impression with their spiky black coiffure, yellow eyebrows and bright red eyes. The most timid are the gentoos, small black birds with white features on their heads; they are the only species to live ashore at Macquarie Island throughout the year and to nest in grassy mounds.

Macquarie Island is home to four species of albatross: wandering, black-browed, grey-headed and light-mantled sooty. The island's fragile population

Opposite: A colony of rockhopper penguins, aptly named for their incredible ability to climb what seem like vertical faces, just near the main station.

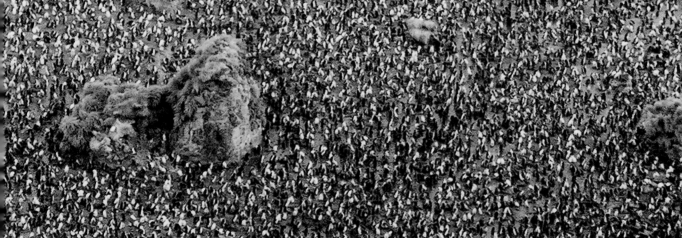

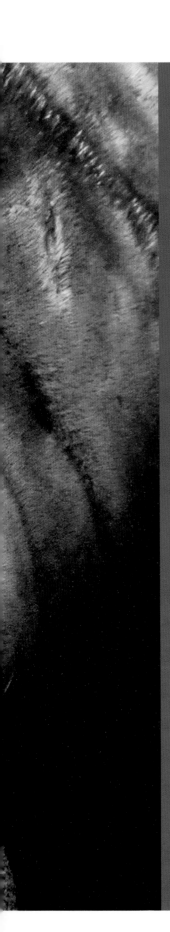

The vanishing

The southern elephant seal and subantarctic fur seal are each listed as 'endangered' under Tasmanian legislation and 'vulnerable' under Australian law, meaning that they are judged to be at risk of extinction in the medium-term future. Threats to the animals include competition and interaction with legal and illegal fisheries, pollution, disease, climate change and human disturbance. Southern-elephant-seal populations are estimated to be declining at 2.5 per cent per year, while the situation for sub-Antarctic fur seals is complicated by interbreeding between this species, Antarctic fur seals and New Zealand fur seals. All four albatross species breeding on Macquarie Island are recorded as threatened, along with several petrels, the Macquarie Island shag and a subspecies of the fairy prion.

But the present regimes for the protection of threatened species came too late for at least three species: a fur seal (destroyed in the island's first decade of commercial sealing), a flightless land rail and a parakeet. The rail (*Gallirallus philippensis macquariensis*) and the parakeet (*Cyanoramphus novaezelandiae erythrotis*), which are close relatives of species that are relatively common in other parts of the southern hemisphere, survived the early introduction of the cat to Macquarie Island, but were all gone within ten years of the introduction of rabbits and wekas. The weka, an aggressive bird, is assumed to have actively contributed to these extinctions, while the rabbits assisted by providing a year-round food source for cats, thus increasing the numbers of these super-predators.

This elephant seal bull lies exhausted following numerous thundering battles to retain the dominance of his harem.

'NOWHERE DOES VEGETATION TOWER; EVERYWHERE IT CLINGS LIKE A CLOSE-CROPPED PELT'

of wandering albatrosses plummeted to a low of just two breeding pairs in 1982, largely as a result of incidental deaths associated with longline fishery in the Southern Ocean. Tens of thousands of sea birds are killed in this manner each year in the southern hemisphere alone. Restricted human activity in the south of the island—where the majority of the wanderers breed—has contributed to a gradual recovery.

+ + +

Macquarie Island, noted Douglas Mawson, 'is devoid of anything in the nature of trees'. The island's plant communities range from the mosses, lichens and dwarf flowering species that form the green bands on the feldmarked peaks, to the marshy 'featherbed' of the lowlands. But nowhere does vegetation tower; everywhere it clings like a close-cropped pelt.

The island's tallest plant is a tussock grass, *Poa foliosa*, which grows in dense clumps on the island's slopes and sections of the coastal

Opposite: A skua hovers watchfully over the penguins gathered below. Forever on the hunt for the weak or distracted, a skua will seise any opportunity for an attack.

terraces and highland valleys. *Poa foliosa* borders the beaches—the hollows between grassy clumps provide convenient wallows for moulting elephant seals—and in places even ventures out onto the sand. At elevation, small nesting birds make their homes beneath *Poa foliosa*'s sheltering blades.

The plants of Macquarie Island are closely related to those found on neighbouring subantarctic islands and species diversity is relatively low. Just three species of endemic plant have been identified: a cushion plant (*Azorella macquariensis*), a grass (*Puccinellia macquariensis*) and the world's southernmost orchid (*Corybas dienmus*). But the glamour twins of the Macquarie Island plant world are the Macquarie Island cabbage (*Stilbocarpa polaris*) and the silver leaf daisy (*Pleurophyllum hookeri*). Throughout the year, the soft and sage-green leaves of the silver leaf daisy provide a luxuriant carpet across the highland herb fields of the island: a display that intensifies in early summer when the plant flowers.

With its robust, rhubarb-like stalks, fans of bright green and thickly haired leaves, and bright bursts of yellow flowers, the abundant cabbage plant was found by early visitors to Macquarie Island to be useful in warding off scurvy. The taste of the

Opposite: Royal penguin colony, hundreds of metres inland from the coast.

plant has been described both as 'a hairy celery' and 'something like that of a cabbage stump'. Mawson recorded that its roots and stalks could be made into a passable soup if they were scraped, cut thinly and boiled. While such soup, he said, was the fare of the workers, the officers' plates were sometimes graced with 'pickles made from the roots'.

While humans have never eaten Macquarie Island plants in any great quantity, the same cannot be said for rabbits. The grazing habits of these introduced animals have, over a century and a half, altered the island's vegetation communities by creating conditions favourable to some plant species and unfavourable to others. In recent years, overgrazing by the booming rabbit population on the island has led to wide-scale erosion and landslips that have destroyed bird burrows and nesting sites. Although work has begun on eradicating the estimated 130 000 rabbits currently grazing on the island, it is not yet known whether all plant species will be able to successfully recolonise in the worst-affected areas.

Opposite: Early-morning dew tells that the wind eased a while and that moments of peace are possible in such a wild place.

In the subantarctic plant house at Hobart's Royal Tasmanian Botanical Gardens, botanists have come up against the challenges of growing Macquarie Island plants some 1500 kilometres north of their natural home. A heavily insulated building provides the plants with near-freezing temperatures and constantly foggy conditions, but wind is also crucial in fostering healthy subantarctic specimens. Constant buffeting, rather than damaging the plants, helps them to develop strength and robustness: a kind of resistance training. In the absence of wind, botanists have sometimes employed the services of an industrial carpet drier to give their charges a decent fitness workout.

Opposite: Flower heads of the Macquarie Island cabbage at 'Sanity Falls'—a name I gave to the place where I sometimes came to recuperate.

Hurd Point, the southernmost
accessible point of the island.
In the centre, below the
escarpment, is the sea of
royal penguins.

THE LAST IMPORTANT STRONGHOLD

Tooth, beak and claw

When sealing and oiling ceased on Macquarie Island in the early years of the twentieth century, devastated populations of seals and penguins were not all they left in their wake. Like splinters of glass in the island's future were populations of feral cats, rabbits, rats, mice and flightless weka.

Cats came to the island in the very first years of sealing, and had established a feral population by the time Russian scientist Bellinghausen visited in 1820. Since the island was virtually unvisited between the 1830s and 1870s, little is known about cat activity in these decades. It is estimated that the cat population quickly built into the hundreds— a death force capable of killing 60 000 birds each year. Some species of burrow-nesting birds survived only by building their nests on offshore sea-stacks inaccessible to feline invaders. A weka-and-cat-eradication program began in 1985 and wekas were completely removed within three years. Efforts on the cat front intensified in the late 1990s and by 2002 Macquarie Island was declared cat free.

The next challenge to confront the island's custodians is the eradication of rats, rabbits and mice. Rabbits, first released by sealers in the 1870s, were a century later numbering in the vicinity of 150 000. Myxomatosis was introduced as a control measure in 1978 and, in 2002, the estimated rabbit population was as low as 16 000. Since then, numbers are calculated to have increased eight-fold and overgrazing has led to large-scale erosion. Factors believed to contribute to this growth include rabbits developing a resistance to myxomatosis, and climate change providing warmer, drier winters that enable rabbits to breed successfully throughout the year. Passionate advocates for the island encouraged the Tasmanian and Commonwealth governments to commit to an ambitious $24.6-million plan to rid the island of its remaining pests. But, as the island's current management plan notes, the 'rugged topography of Macquarie Island, together with the rodents' widespread abundance and reproductive capability mean that eradication will be a considerable challenge'.

'the most wretched place of involuntary and slavish exilium...'

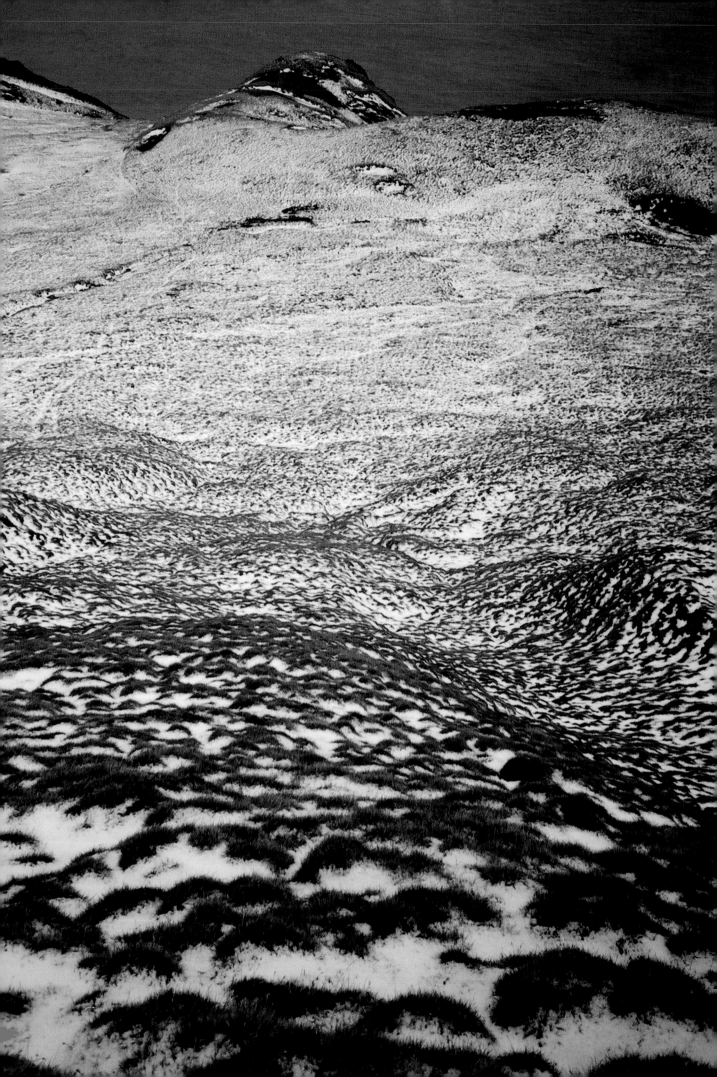

'LIVING CONDITIONS FOR THE SEALERS AND OILERS WERE AS BRUTAL AS THE TASKS THEY DAILY UNDERTOOK'

In the human imagination, islands have forever been bounded by the concepts of prison and paradise; in two centuries of human history, each of these labels has been affixed to Macquarie Island. Although never a penal settlement—being deemed too remote, stormy and sterile even for that grim purpose—it was for a century a place of inescapable hardship for the men (and the occasional woman) who lived, laboured and died on its shores. The ships that landed their gangs of sealers and oilers on the remote speck of land were themselves controlled by the mercurial forces of distant commerce and foul weather. Ever present for the work crews was the risk of being left on the island well beyond the exhaustion of food and fuel supplies. Even when provisions did hold out, living conditions for the sealers and oilers were as brutal as the tasks they daily undertook.

Macquarie Island's recorded history began in the winter of 1810 with the arrival of the brig *Perseverance*, a Sydney sealing ship under the command of Captain Frederick Hasselborough. The 'discovery' of the island was soon followed by the revelation of an earlier human connection to the island. The wreckage of a ship, much later described by Douglas Mawson as being of 'ancient design', was found high in the tussock grass on

Opposite: It snows even in summer on Macquarie Island, though with the average temperature of 4°C, it doesn't settle for long.

the wind-whipped west coast. The origins of the ship and the precise destiny of its crew remain a mystery. The wreckage seems to have disappeared, most probably into the fires of those early work gangs who tried to eke out a living on the bitterly cold and treeless island.

It was Hasselborough who named the island, for Governor Lachlan Macquarie, before moving quickly to capitalise on the wealth of fur seals he saw clustered thickly on rocks and beaches. Landing a gang of eight workers with several months' provisions and instructions to get to work, Hasselborough set sail for Sydney, where he intended to employ more sealers, stock up on salt for curing skins and keep secret the location of his newly discovered resource. He succeeded in the all but the last of these missions. By the end of 1810, Macquarie Island's latitude and longitude were well known, and commercial-scale plunder by several sealing gangs had begun.

The 1813 log of the brig *Mary and Sally*—just one of a multitude of ships whose crews were working the island—paints a bloody picture of the helpless seals and their ruthless predators: 'At 11am the boat was sent to shore to look for seal. At 2pm the boat returned with 256 skins and was hoisted in'. The killing was so relentless that,

Opposite: Control valve on a steam digestor chamber. The chamber is rusted but the hard-wearing brass valve is a reminder of the extent of industrialisation in the early years.

'THE KILLING WAS SO RELENTLESS THAT, ELEVEN YEARS LATER, ONE SPECIES OF FUR SEAL HAD BEEN ELIMINATED FROM THE ISLAND'

eleven years later, one species of fur seal had been eliminated from the island and stocks of the remaining species had been so depleted that sealers managed to take only four skins in an entire season. The fur seal population spent, the sealers' attention turned to elephant seals, the blubber of these abundant animals producing richly marketable oil. By 1830, the island's elephant seal population too had been devastated. The entry in the *Mary and Sally* log for 18 January 1832 records that the crew were 'getting things aboard ready to move to Ballast Bay as the elephant is all killed at the North End of the Island'. In that year, the harvest ceased: there was nothing left worth taking.

Macquarie Island's next four decades were almost entirely empty of human visitation. While try-pots rusted at the high-water mark, while mud and tussock huts dissolved into the hinterland and while shelter cobbled together out of shipwreck timbers were blown apart by the Furious Fifties, the fur and elephant seals of Macquarie Island staged a slow and partial comeback. Then came the island's second era of exploitation, an era dominated by a single man.

Joseph Hatch was short, quick, ruthless, teetotal and unswervingly controversial. A chemist and small-time statesman in New Zealand's south,

Following pages: An elephant seal pup separated from its mother by a bulldozing beach master (the dominant bull). It lies beside the skua-pecked carcass of another pup squashed by the bull's rampage.

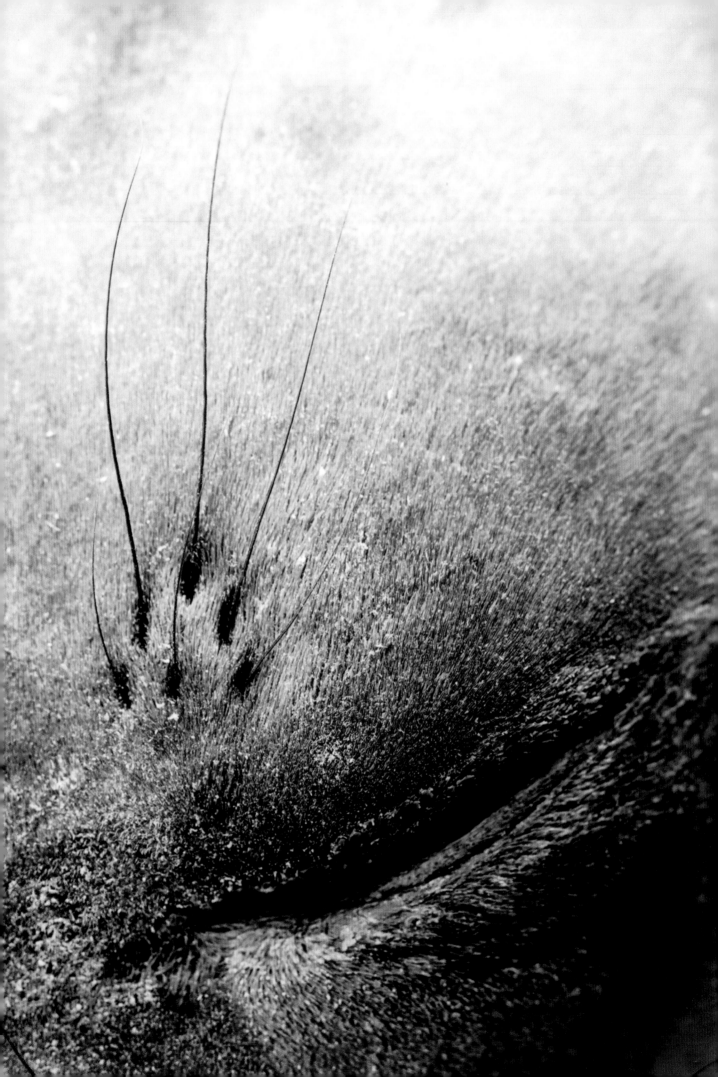

he doggedly pursued various entrepreneurial projects throughout his career, never quite backing the right horse. Bone-milling, sheep-dipping, poisoning grain for pest control, soap-making and exporting rabbit skins were among his endeavours, but it was his determination to grind a fortune out of the wildlife of a tiny subantarctic island that was to consume thirty years' worth of his copious energy.

After his parliamentary career was scuppered by allegations that his employees had been poaching seals in New Zealand's closed season, Hatch, in 1887, turned his attention to the revived elephant seal industry on the remote Macquarie Island. While Hatch would tout his 'Sea Elephant Harness Oil' as 'the best unadulterated oil for dressing leather in the world', he soon saw that the trade in elephant seal oil alone would be insufficient to make him a wealthy man. Seal numbers were plummeting again in the face of renewed commercial interest and the remaining animals, although easy to shoot and kill, could not be relied upon to come ashore for capture close to the places where they could be processed. Hatch's men took to flensing seals where they were shot and carrying their seeping blubber back to the oiling plant on their backs.

Opposite: A female elephant seal, resting.

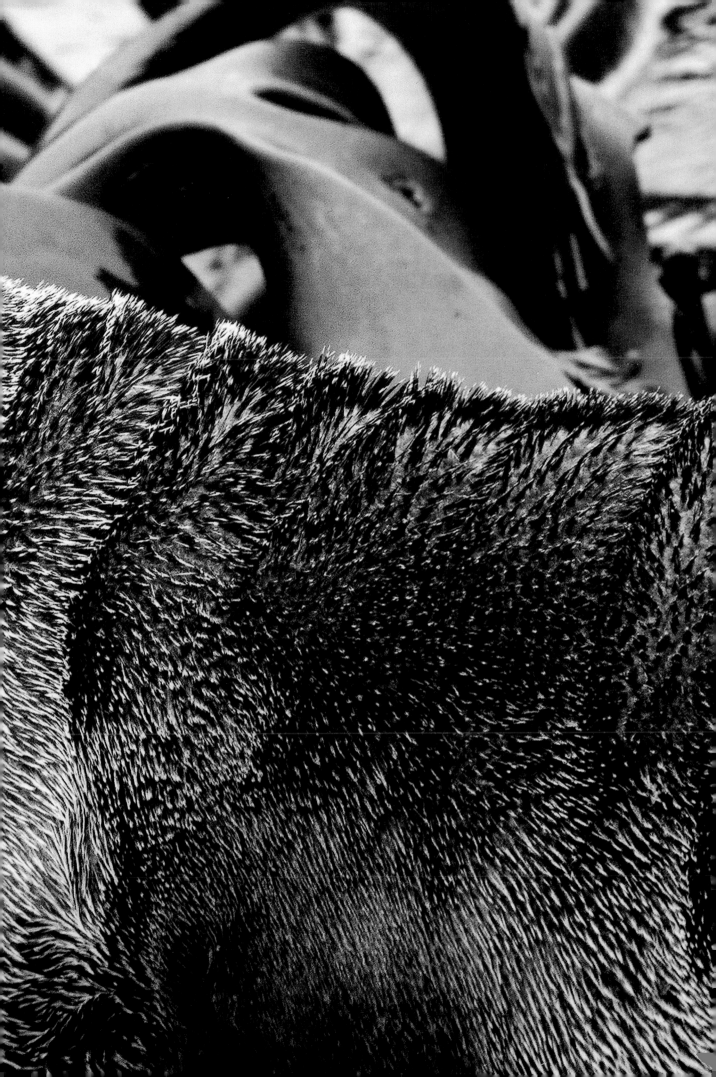

It would be penguins—their oil less valuable than that from elephant seals but still useful in rope and twine production—that would form the backbone of Hatch's industry and fuel his aptly named Southern Isles Exploitation Company. Penguins were not only defenceless and plentiful, they also had the convenient habit of massing in predictable places. New steam digester technology, allowing oil to be extracted from animal bone and skin as well as blubber, made the previously low-yield penguins worth pursuing. In 1890 Hatch installed a steam digester plant in the king penguin rookery at Lusitania Bay in Macquarie Island's south. In the years that followed, he expanded, setting up plants at strategic points across the island. As king penguin numbers diminished, Hatch moved on to the smaller royal penguins, concentrating his efforts at The Nuggets in the island's north-east.

An employee of Hatch's provides a graphic description of the industry in those years before the turn of the century:

> the penguins, when coming from sea, can be easily driven and yarded like sheep. Where the yard is full, 10 men go out and club the birds before breakfast. When work is resumed many of the poor birds are found to have recovered and are walking about; they required

Opposite: Juvenile elephant seals huddle together to conserve energy. They will wait many years for their opportunity to reproduce.

'BY 1909, THE DIGESTERS AT THE NUGGETS WERE CONSUMING AN ESTIMATED 3500 ROYAL PENGUINS A DAY'

reclubbing . . . As many as 2000 birds can be put through the digesters in a day, equal to 14 casks of oil, each about 40 gallons [182 litres].

By 1909, the digesters at The Nuggets were consuming an estimated 3500 royal penguins a day. It is hard to know whether Hatch truly believed his own rhetoric when he claimed, in a prospectus designed to shore up public support for his industry, 'it is impossible—if the quantity of Oil now taken were doubled—to materially reduce the number of birds'.

By the turn of the century, Macquarie Island's remoteness no longer entirely veiled Hatch's enterprises from public view. In 1911, five men from Douglas Mawson's Australasian Antarctic Expedition landed on the island with a program of scientific work outlined by their leader. Stories and images generated by the heroic era of Antarctic exploration had already kindled a worldwide love affair with the penguin, a fascination that to this day shows no sign of growing dim. As the jaunty birds took up their tuxedoed places in popular culture, pressure began to mount on the Tasmanian government to put an end to Hatch's slaughter. Meanwhile petroleum products were swiftly reducing the demand for animal oil. The

Opposite: This silhouette is more benign than it looks. There are no toilets at coastal field huts and so approved practice is to send it out to sea with all the other animals' waste.

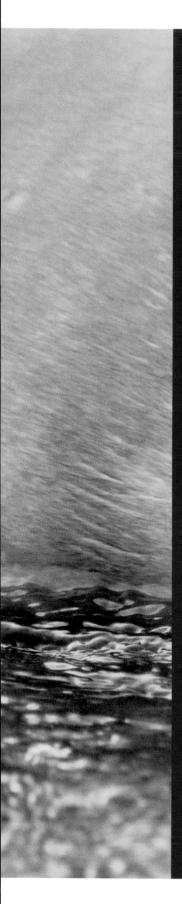

Mawson's plea

8th December, 1932

The Chief Secretary,
Government Offices,
HOBART

Dear Sir,
It has come to my knowledge that some sections of Tasmanian
folks are making representation to the Government to ensure the
continuance of Macquarie Island as a sanctuary for the bird and seal
life of the subantarctic.

I send this note to express my conviction that this is the best possible
use to which the island can be put.

If such can be arranged, Tasmania will always benefit from the
esteem of people all over the world. If, on the other hand, slaughter is
licensed, the immediate reward to the Tasmanian Government would
be very small and it would not be long before all the assets would
have disappeared, and the island would have no future whatever.

Believe me,

Yours very truly,
DOUGLAS MAWSON

plight of the Macquarie Island penguins made headlines in newspapers across the Western world. Finally, in 1919, Hatch's licence to take wildlife on Macquarie Island was revoked. It was 109 years after Hasselborough had landed his small party of sealers that the island made its return from wildlife resource to wildlife refuge. Mawson was among those who saw off later attempts to reopen a Macquarie Island sealing season, and who agitated for the legal protection that was finally to be bestowed upon the island in 1933.

During Hatch's fight to keep his penguin-oil industry alive, he once observed: 'The Macquarie Island oil industry carried on by myself for a number of years, has been fraught from time to time with such vicissitudes that many, even the most enterprising men, would long ago have abandoned it'. This bloody-minded determination to succeed against the odds, over Hatch's thirty-year connection with the island, resulted in an estimated three million Macquarie Island penguins being rendered into oil. Today the rusting and robotic remains of Hatch's digesters are the most obvious signs of the island's century of bloodshed. They are regarded with a complete lack of interest by the residents of the thriving penguin colonies that envelop them.

Opposite: King penguins in a show of force at the Lusitania Bay colony.

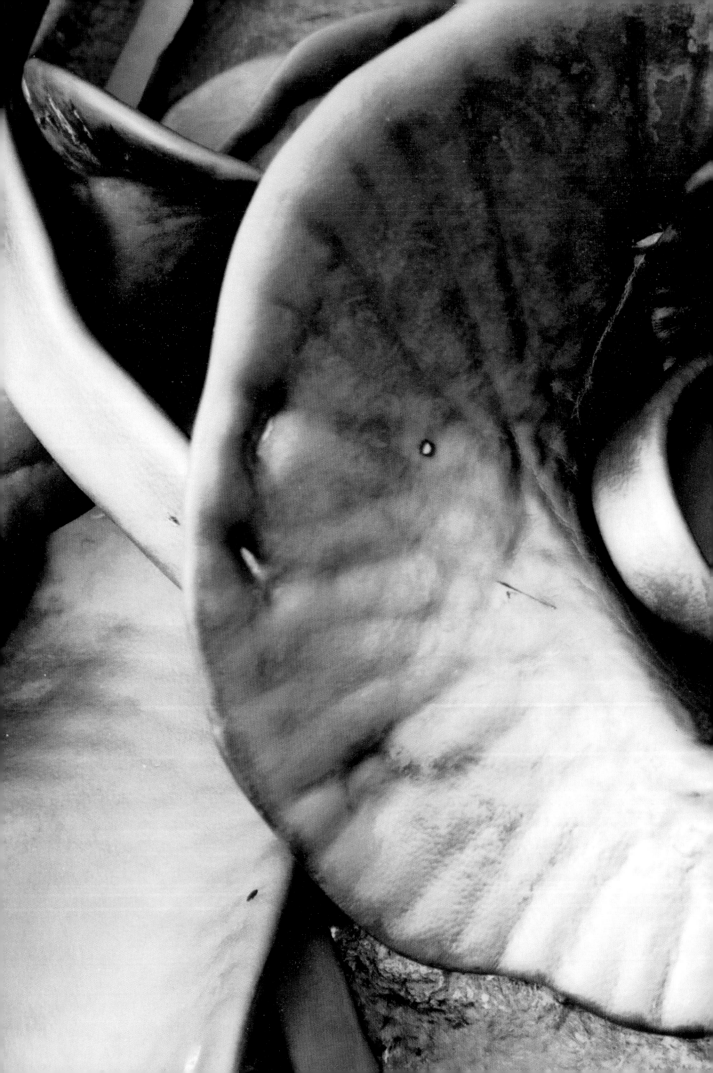

'an exquisite
scene of
primitive
nature ...'

Macquarie Island is not a human place. Over time, people have come to its shores by way of shipwreck, or to work, or to visit, but never simply to live. Today, it is precisely this lack of human presence that makes a visit to the island such a sought-after experience. Visitors come eager for a glimpse of a world beyond human civilisation.

Human activity on the island is concentrated at the north end, where low buildings and demount-able huts cluster on the isthmus in a tiny frontier township. The main street, demarcated by the wheel tracks of a handful of vehicles, is some-times occupied by lounging elephant seals. Through the winter months, a skeleton crew of rangers, scientists and support workers occupies the island. Then spring brings ships from the north, swelling the island's human population to its summer proportions with more expeditioners and with tourists prepared to pay handsomely for the small glimpses they are afforded of the island and its thronging wildlife.

Perhaps those with the closest ties to the island are the repeat expeditioners whose work over the years has involved monitoring the island's features, species and communities, both terres-trial and marine. Theirs is work that takes them beyond the confines of the northern isthmus and

Opposite: Brothers Point 'Googie' hut. Even the human infrastructure seems to take on animal features.

its station, and into the heart and the south of the island. Here, they say, they come as close as possible to escaping civilisation. One thread that runs through the experiences of these women and men is the paradoxical sense they have had of being at once far from home and more at home than ever before. Another is their determination to work for the protection of the island and its ecosystem, which is still under threat from human actions, both past and present.

Rough, wild places are virtually in the DNA of ecologist Jenny Scott. The cherished places she has worked—Scotland, Iceland, subantarctic Hurd Island, south-west Tasmania and Macquarie Island—are all indelibly printed on her sense of self. Since her first visit to Macquarie Island in 1979, Scott has returned time and time again, working on various projects including monitoring vegetation communities and albatross populations.

'My longest visit was sixteen months, and my shortest four days. I have a tremendous sense of belonging when I'm there. It is a special thing when you know a place so well that you know what all the plants are, you know what all the animals are, you know all the predicted behaviours and you know all the seasonal,

Opposite: Green Gorge Hut at night—darts, fortified wine, books and conversation.

Joseph Burton: a 'touched' taxidermist

Not all visitors during the age of exploitation were insensible to the beauty of Macquarie Island and its remarkable abundance of wildlife. Writing about his three-and-a-half-year tenure on the island in the closing days of the nineteenth century, collector and taxidermist Joseph Burton observed: 'the island is dreadfully dreary to the ordinary observer, but to the naturalist it is full of fascinating interest'.

Burton, who worked at New Zealand's Colonial Museum in Wellington during the late 1870s, was employed on Macquarie Island between November 1896 and April 1900 by oil-industry entrepreneur Joseph Hatch. He was allowed by Hatch, on his days off, to collect sea-bird eggs for A J Campbell, author of *Nests and Eggs of Australian Birds*. He was also responsible for capturing a pair of king penguins that became live exhibits at the Melbourne Aquarium.

'During my spare time I fossicked everywhere', Burton reported. 'Some of my hunts led to my being benighted in exceedingly rough localities. When I used to sit for hours on a Sunday watching the birds and seals I guessed by the askance looks of my mates they thought I was a bit "touched".'

Burton observed and recorded some extraordinary sights, such as leopard seals 'planting' in heavy kelp to 'waylay unsuspecting penguins' as they came ashore. The seal, Burton reported, 'makes a dextrous spring and usually grabs the bird, giving it a jerk with such force as to tear a mouthful out; the bird is thus devoured in mouthfuls, feathers and all'. He also recorded observing skuas ('a murderous creature, worse than any hawk ever hatched') playing ball in the air with a king penguin egg: 'Every time the egg was let go it was caught by another bird, and so the fight continued, the egg being released and caught several times before it dropped on the sandy beach and was broken to bits.'

Although contemporary conservationists would be horrified by some of his activities, such as visiting penguin rookeries with his dog and allowing it to clear a path through nesting birds by pushing them this way and that, Burton was their forebear in that he was able to see the island's natural workings and wonders.

This royal penguin, having just (barely) survived a leopard seal attack, fell prey only a few hours later to a giant petrel.

'IT'S AN INTENSE FEELING
THAT I HAVE WHEN I'M THERE'

cyclic patterns. Macquarie Island is a roaring, wild place. Nevertheless, you feel very at home there. It's an intense feeling that I have when I'm there, and a very privileged one', she says.

'Perhaps the epitome of a Macquarie Island experience is to stand on Petrel Peak on a wild, stormy day: one of those days when the seas are huge and yet the sun is shining. Squalls come over, and then they're gone. There's no-one else for miles and miles and you are totally at one with your surroundings.'

Working in Antarctic tourism, she says, has enabled her to keep fresh her feelings for the island. At the same time, her intermittent visits over the past decade have made her an important witness to the devastating changes wrought in recent years by rabbits on the island's southern slopes.

Macquarie Island's rabbit population has now risen back to levels not seen since before the introduction of myxomatosis in the 1970s. The rampant grazing of tussock grass and cabbage plants on steep slopes is causing grand-scale erosion, leading in turn to the destruction of long-established albatross nests and the loss of shelter for small, burrow-nesting birds.

Previous pages, left to right: Ornithologist Nick Holmes holding up the staple mango chutney, sardine, cheese and cracker lunch; Me at Sanity Falls, taking a few moments.

Opposite: Nick enjoying an extremely rare clement day, in the herb field above Sandy Bay.

'THE ISLAND REMOVES
ME FROM CIVILISATION'

Albatross biologist Aleks Terauds describes Caroline Cove in the island's far south as his 'first real home'. All told, he spent three years in this place, keeping watch over the island's precarious albatross populations, tracking the slow and still uncertain increases in their numbers. Data collected over decades of such monitoring has been important in making the link between dwindling albatross populations and longline fishing in the Southern Ocean.

'The island is elemental. Raw. While it's not pristine—there *are* human impacts—it hasn't been perverted by people', he says. 'The island removes me from civilisation. It's one of those rare places where you can still be in touch with nature without having to first break through all the barriers that civilisation erects between humans and the Earth.'

For Terauds, that the island has for all time been completely separate from any other landmass is an essential part of its character. In his mind, it is a floating ark: inseparable both from the ocean that surrounds it and the weather that hammers it. Like all those who have fallen in love with Macquarie Island, Terauds is fiercely protective of the place. He is especially impatient with the homocentric attitude that the island and its

Opposite: King penguins taking the opportunity to rest and recuperate at 'The Public Baths'—a spot set back from Sandy Bay.

Following pages: The light-mantled sooty albatross.

wildlife exist for the viewing pleasure of visitors. 'We must remember that this is a wild place, not a zoo.'

+ + +

As soon as she set foot for the first time on a Macquarie Island beach, artist Catherine Bone was prepared to commit. 'If someone had appeared in that moment, and said "Choose, right now, whether or not you want to stay here for the next ten years", I would have said "Absolutely, I'll stay". Because that's how Macquarie Island makes you feel—like you never want to be anywhere else.'

Bone had come from the Australian tropics to the island's subantarctic climate and the numerous canvases she produced during her stay represent a dedicated flirtation with hypothermia. Bone would sit, for hours at a stretch, on the fringes of seal and penguin colonies, brush in hand, recording the island's features in spite of her increasing coldness.

'The thing that struck me about Macquarie was its unity. It seemed to me that the animals were part of the landscape. They *were* the landscape. Everything was so perfectly connected. And you, too, become more of an animal than a person

Opposite: Bauer Bay, Sandy Bay and Overland Track junction: timber signs worn from the relentless wind and sand.

'AND YOU, TOO, BECOME MORE OF AN ANIMAL THAN A PERSON WHEN YOU'RE THERE. YOU'RE PART OF THE LANDSCAPE, YOU'RE PART OF THE ANIMAL WORLD.'

when you're there. You're part of the landscape, you're part of the animal world. You feel right, not separate from everything the way you are in this civilisation.'

Bone's partner is wildlife biologist Nigel Brothers, one of the world's foremost protectors of albatrosses and other sea birds. Brothers is another who has made multiple journeys to the island, and years' worth of his detailed, handwritten Macquarie Island field notes are held by the State Library of Tasmania.

'The island is always changing—there's something completely new around every corner. That is part of its excitement', he says.

Ever the pragmatist, Brothers is also quick to point out that the contemporary 'homeliness' of the island is, in part, the result of easy access to creature comforts.

'One hundred years ago, people were describing this island as the worst place on the planet, and perhaps part of the reason for that, and why it is so much less threatening to us now, is because the base is there, the field huts are well established, and we are able to be there in safety and comfort. Once, people died there. They just perished up on the plateau.'

Opposite: Nick analysing data from a day at the royal penguin colonies. In summer, daylight extends for twenty hours.

Another pragmatist, retired wildlife management officer Geof Copson, calculates his Macquarie Island tenure—in numerous visits between 1974 and 2006—at six-and-a-half years. It is probable that only some of the workers who occupied the island during the Joseph Hatch era spent longer on the island. Despite his intimate knowledge of the place, and deep concern for its future, Copson is not one for romanticising.

'I'm one of those funny types who don't get particularly attached to place. I liked the work, it was very worthwhile. And I was as far away as I could get from head office. In the early days I was there, there wasn't much in the way of communication, like there is now. That's probably the thing that's changed most, and to be honest, I preferred it how it was,' he says.

Once star-remote, Macquarie Island is today closer to the rest of the world than ever before, woven into the global community by the bright strands of electronic communication and sophisticated shipping. Web-cam can show you the scene at the isthmus at any moment of any day. Although this extraordinary island may seem to exist on the fringes of our imagination, it is increasingly bound to our material reality.

Opposite: Rockhopper enjoying a moment of sunshine.

Commenting on the future of the island, Copson lists global warming, pollution and increased fishing activity in the Southern Ocean as just some of the threatening processes that will continue to affect the flora and fauna of the place Sir Douglas Mawson once called 'one of the wonder spots of the world'. His words remind us that neither the island's manifold legislative protections nor its geographical remoteness can put it beyond the consequences of human choice.

Opposite: Friend and fellow researcher, Brian Reis, experiencing the inquisitive nature of king penguins at Green Gorge beach.

Macquarie Island 2056

In its 2006 Management Plan for Macquarie Island, the Tasmanian Parks and Wildlife Service sets out its vision for Macquarie Island fifty years into the future:

Macquarie Island is a nature reserve where all of the World Heritage values, Biosphere Reserve values, National Estate values and state nature reserve values are protected and conserved. There is a relatively unaltered natural diversity, including geodiversity and biodiversity. The populations of some threatened species in the reserve appear to be recovering, even if their populations are still threatened elsewhere. Human visitation and use of the reserve is controlled and carefully managed to minimise adverse impacts on the reserve. Scientific research, monitoring and management programs continue with minimal and/or transitory impacts on the natural and historical values of the reserve. There have been no apparent further introductions of alien species. Rabbits, rats and mice have been eradicated. There is full awareness and appreciation of the special conservation value and character of the reserve by the international community, the Australian federal, state and local governments, scientists, tourists and the Australian public, to the extent that protection of the reserve is recognised to be of utmost importance.

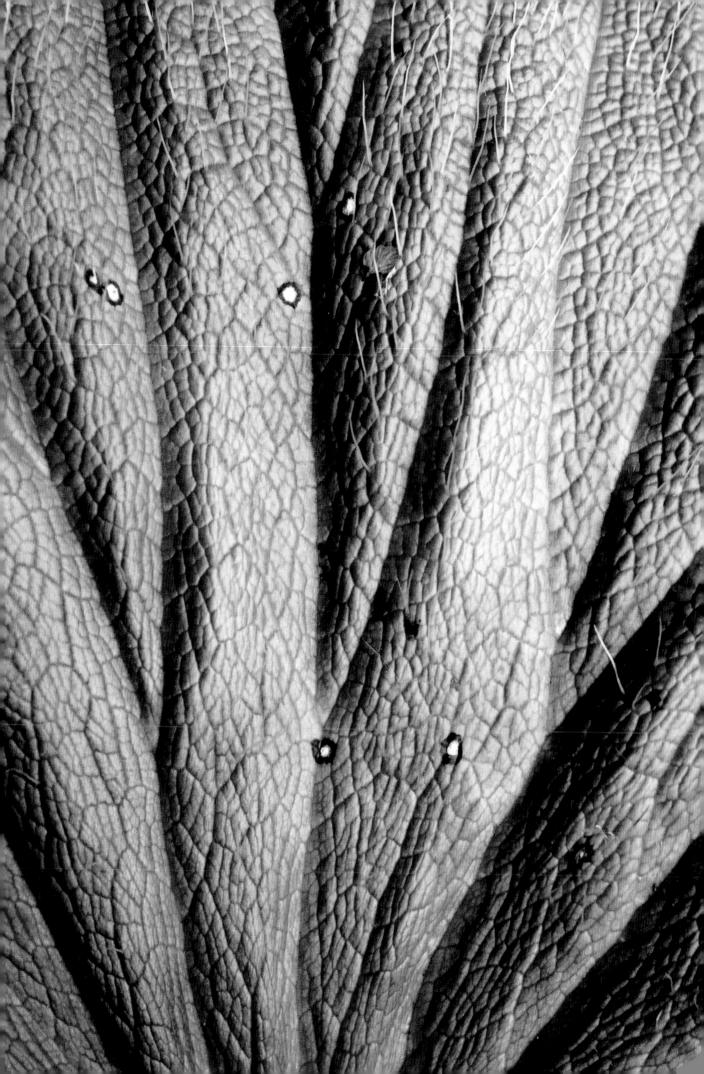

AFTERWORD

Turn through the pages of this book again and try to imagine that you are sitting at the top of a mountain, over five-and-a-half kilometres high in the sky.

Now, if you can, imagine the most beautiful ocean, wild and clean, and the sounds and smells of ancient sea creatures.

Picture in your mind a long, narrow strip of rocky land; it's the summit of this 5500-metre mountain jutting straight out of this pure and powerful ocean. This is Macquarie Island, the top of one of the world's tallest mountains, with only the upper few hundred metres above sea level.

Those who have made it to the tops of mountains of some description and size always speak of the freedom they feel when they are up there, and the pure exhilaration of being so at one with nature.

For me what makes Macquarie Island unique is its mixture of spectacularly diverse and abundant wildlife, the unique flora and climate, where it sits on our planet and its proximity to the freezing Antarctic waters, but most of all, it is the energy of the island, the spirit of the island.

Opposite: A blissful smile. Fat from mother's milk and nestled within the crèche, the weaner is yet unaware of the harshness to come.

The energy or the feeling that people get from being at Macquarie Island is one of a kind. My theory is that it's the powerful combination of two pure spaces, two spaces that as humans we resonate with: the summit of a mountain and the ocean.

Energetically pure natural landscapes.

Add the smells, sounds and sights of all the inhabitants of Macquarie Island, and you experience something unique to our galaxy, not just our planet.

This is wilderness: it is raw, it is beautiful, and it is hostile.

The animals that inhabit this landscape are resilient, but not immortal. They have established populations, family groups, life-long partners, burrows as their homes, and they have survived all imaginable threats—slaughter being the most obvious and controversial.

But loss of habitat is the current threat affecting species of plant and animal at Macquarie Island. Early industry on Macquarie Island introduced

brought with them soil, seeds, fire, foreign animals and a thirst for money; all of these did not and do not belong in this landscape. Of these imports, cats and wekas have been eradicated. Remaining are the rabbits and rodents.

A rabbit-and rodent-eradication program has been in the planning for many years, and the first attempt at an aerial-baiting program was made in 2010. With ongoing support from the federal and Tasmanian governments, this program will begin in the winter of 2011.

Rabbits and rodents pose a severe threat to the values of Macquarie Island. Ongoing research has revealed widespread damage to ecosystems, including destruction of vegetation and catastrophic erosion, dramatically affecting numerous species of flora and fauna.

I am inspired by the commitment of the project officers working on this program, and all others who have had a positive influence along the way. The personal commitment and perseverance of these people reminds me that we have come a long way in appreciating and protecting the wonders of our planet.

We have much to learn and experience as a species, as we continue to expand our own global population. Evidence about climate change is now black and white; it is what we do as a global community with this knowledge, awareness and hope that will be the difference between retaining jewels like Macquarie Island, and modifying all landscapes and ecosystems to such a degree that we remove all pure places of sanity.

Macquarie Island, with its simple and hostile beauty, can teach us so much. In my way, I have tried to bring this message to you.

Alistair Dermer

Previous pages:
Silver-leaf-daisy field on a western coastal slope. On my return visit in 2009 this area was decimated by the impact of rabbits.

ACKNOWLEDGEMENTS

I would like to thank a number of people for a number of things; below is a list of those who have helped me in one way or another with this book.

Thanks to Bob Brown for his guidance and encouragement; Pat Sabine for her faith and support that made the first steps of this book possible; Nick Holmes for his friendship and patience, undertaking his PhD at Macquarie Island, and needing a field assistant; Mel and Brian for their friendship and belief in my abilities; Jayne and Beverly for their early involvement; and to my friends and who have provided either a home, an introduction or just a kind word of support at the right time.

Thanks to Danielle Wood for her contribution, to Lily Keil for her gentle and professional approach in getting the book finished, Tracy O'Shaughnessy for taking it on in the first place and Foong Ling Kong for the title and belief. It has been a long time in the making, and all have contributed the right amount at the right time; thank you.

Thanks to my parents for their balanced guidance, creative awareness and acceptance and support of my dreams and aspirations.

And a special thanks to my best friend, wife and soul dancer, Karen, for her love, patience, support and encouragement.

SOURCES

'like a signpost pointing the way to the frozen
lands . . .'
Macquarie Island chronicler J S Cumpston, 1968,
in his book *Macquarie Island*.

'the last important stronghold of subantarctic
bird and other animal life . . .'
Activist for the conservation of Macquarie Island
wildlife William Crowther, 1933, in an article for
Emu magazine.

'the most wretched place of involuntary and
slavish exilium . . .'
Captain Douglass of the ship *Mariner*, 1822,
quoted in Robert Cox's pamphlet 'One of the
Wonder Spots of the World'.

'an exquisite scene of primitive nature . . .'
Antarctic explorer Douglas Mawson, 1914, in his
book *The Home of the Blizzard*.

THIS BOOK WAS DESIGNED AND
TYPESET BY PFISTERER+FREEMAN
THE TEXT WAS SET IN 9 POINT CAECILIA
WITH 7 POINTS OF LEADING
THE TEXT WAS PRINTED ON 130 GSM MATT ART
THIS BOOK WAS EDITED BY SUSAN KEOGH